General Directions

There are several basic techniques and information on materials that will help you in creating our sachets. We have included this information here. We suggest that you read through these General Directions carefully before beginning your project.

Crochet Materials

Steel Hooks:

If you have never worked with steel hooks and finer thread, don't be afraid to try. You will be using exactly the same stitches you're familiar with, but at first it will feel clumsy and awkward. For an experienced crocheter, this is a bit of a surprise—suddenly feeling all thumbs again just as when you first learned to crochet. But this will pass in a few hours of crocheting, as you adjust your tension and working method to the new tools. Soon you will work much more by feel than when working with heavier yarns. So be patient with any initial clumsiness and confusion—they won't be with you long.

Steel hooks range in size from 00 (large) to 14 (very fine), and are 5" long which is shorter than the aluminum or plastic hooks. Their shape is different from the other crochet hooks. There is the throat, then the shank, and after the shank the steel begins to widen again before it reaches the finger grip **(Fig 1)**. When crocheting, it is important that the stitches do not slide beyond the shank as this will cause a loose tension and alter the gauge. If you find you are having difficulty at first, put a piece of cellophane tape around the hook to keep the stitches from sliding past the correct area. With practice, you will work in the right place automatically.

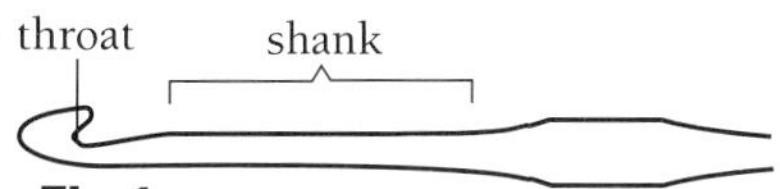

Fig 1

Crochet Cotton:

Bedspread-weight crochet cotton, the size usually used to make tablecloths and bedspreads, is used for the projects in this book. It is sold under various brand names, such as: J. & P. Coats *Knit-Cro-Sheen®* or *Opera*, size 10; Anchor *Crochet Cotton*, size 10; DMC *Cebelia*, size 10; or *Grandma's Best*.

Size and Gauge

A correct stitch gauge is very important. Please take the time to work a stitch gauge swatch about 4" x 4". Measure the swatch. If the number of stitches and rows are fewer than indicated under "Gauge" in the pattern, your hook is too large. Try another swatch with a smaller size hook. If the number of stitches and rows are more than indicated under "Gauge" in the pattern, your hook is too small. Try another swatch with a larger size hook.

Stitch Guide

Chain - ch:
YO, draw through lp on hook.

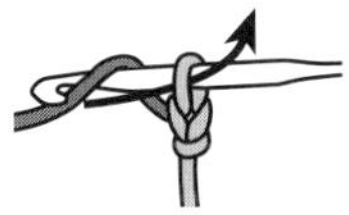

Single Crochet - sc:
Insert hook in st, YO and draw through, YO and draw through both lps on hook.

Reverse Single Crochet - Reverse sc:
Work from left to right, insert hook in sp or st indicated (**a**), draw lp through sp or st - 2 lps on hook (**b**); YO and draw through lps on hook.

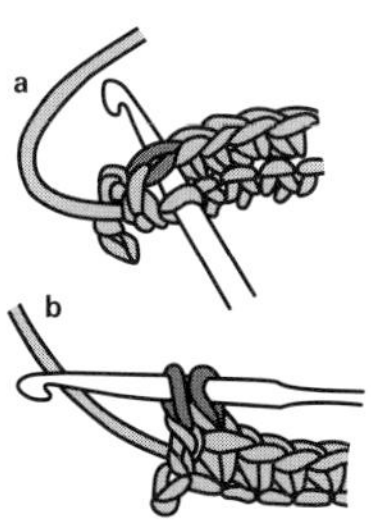

Half Double Crochet - hdc:
YO, insert hook in st, YO, draw through, YO and draw through all 3 lps on hook.

Double Crochet - dc:
YO, insert hook in st, YO, draw through, (YO and draw through 2 lps on hook) twice.

Triple Crochet - trc:
YO twice, insert hook in st, YO, draw through, (YO and draw through 2 lps on hook) 3 times.

Slip Stitch - sl st:
(a) **Used for Joinings**
Insert hook in indicated st, YO and draw through st and lp on hook.

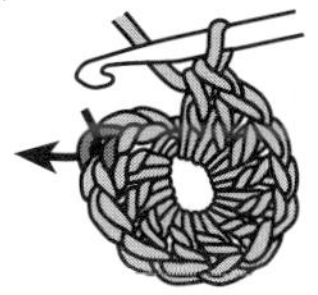

(b) **Used for Moving Yarn Over**
Insert hook in st, YO draw through st and lp on hook.

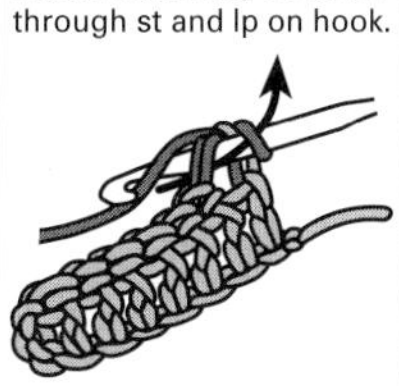

Front Loop - FL:
The front loop is the loop toward you at the top of the stitch.

Back Loop - BL:
The back loop is the loop away from you at the top of the stitch.

Post:
The post is the vertical part of the stitch.

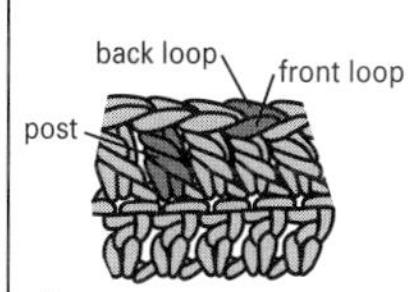

Abbreviations and Symbols

beg . . . . begin(ning)
BL(s) . . . . back loop(s)
ch(s) . . . . chain(s)
CL(s) . . . . cluster(s)
dc . . . . double crochet(s)
dec . . . . decrease(-ing)
dtrc . . . . double triple crochet(s)
FL(s) . . . . front loop(s)
lp(s) . . . . loop(s)
patt . . . . pattern
prev . . . . previous
rem . . . . remain(ing)
rep . . . . repeat(ing)
rnd(s) . . . . round(s)
sc . . . . single crochet(s)
sk . . . . skip
sl . . . . slip
sl st(s) . . . . slip stitch(es)
sp(s) . . . . space(s)
st(s) . . . . stitch(es)
tog . . . . together
trc . . . . triple crochet(s)
YO . . . . yarn over

***** An asterisk is used to mark the beginning of a portion of instructions to be worked more than once; thus, "rep from ***** twice more" means after working the instructions once, repeat the instructions following the asterisk twice more (3 times in all).

† The dagger (or **††** double daggers) identifies a portion of instructions that will be repeated again later in the same row or round.

—The number after a long dash at the end of a row or round indicates the number of stitches you should have when the row or round has been completed. The long dash can also be used to indicate a completed stitch such as a decrease, a shell, or a cluster.

( ) Parentheses are used to enclose instructions which should be worked the exact number of times specified immediately following the parentheses, such as "(2 sc in next dc, sc in next dc) twice." They are also used to set off and clarify a group of stitches that are to be worked all into the same space or stitch such as "(2 dc, ch 1, 2 dc) in corner sp."

[ ] Brackets and **( )** parentheses are used to provide additional information to clarify instructions.

Join - join with a sl st unless otherwise specified.

The patterns in this book are written using United States terminology. Terms which have different English equivalents are noted below.

United States	English
single crochet (sc)	double crochet (dc)
half double crochet (hdc)	half treble (htr)
double crochet (dc)	treble (tr)
triple crochet (trc)	double treble (dtr)
double triple crochet (dtrc)	triple treble (trtr)
skip (sk)	miss
slip stitch (sl st)	slip stitch (ss) or ("single crochet")
gauge	tension
yarn over (YO)	yarn over hook (YOH)
bedspread-weight crochet cotton	lightweight crochet cotton

Metric Conversion Charts

INCHES INTO MILLIMETERS & CENTIMETERS (Rounded off slightly)

inches	mm	cm	inches	cm	inches	cm	inches	cm
1/8	3		5	12.5	21	53.5	38	96.5
1/4	6		5 1/2	14	22	56	39	99
3/8	10	1	6	15	23	58.5	40	101.5
1/2	13	1.3	7	18	24	61	41	104
5/8	15	1.5	8	20.5	25	63.5	42	106.5
3/4	20	2	9	23	26	66	43	109
7/8	22	2.2	10	25.5	27	68.5	44	112
1	25	2.5	11	28	28	71	45	114.5
1 1/4	32	3.2	12	30.5	29	73.5	46	117
1 1/2	38	3.8	13	33	30	76	47	119.5
1 3/4	45	4.5	14	35.5	31	79	48	122
2	50	5	15	38	32	81.5	49	124.5
2 1/2	65	6.5	16	40.5	33	84	50	127
3	75	7.5	17	43	34	86.5		
3 1/2	90	9	18	46	35	89		
4	100	10	19	48.5	36	91.5		
4 1/2	115	11.5	20	51	37	94		

mm - millimeter   cm - centimeter

CROCHET HOOKS CONVERSION CHART

U.S.	1	2	3	4	5	6	7	8	9	10	11	12	13	14
English	3/0	2/0	1/0	1	1 1/2	2	2 1/2	3	4	5	5 1/2	6	6 1/2	7
Continental-mm	3	2.5		2		1.75	1.5	1.25	1	0.75		0.6		

Blue Mood

Size:
About 5 1/2" x 4 1/2"

Materials:
Bedspread-weight crochet cotton, 100 yds blue
Size 6 steel crochet hook, or size required for gauge

Gauge:
8 dc = 1"

Instructions

Front/Back (make 2)
Ch 4.

Rnd 1 (right side)**:**
14 dc in 4th ch from hook (beg 3 skipped chs count as a dc); join in 3rd ch of beg 3 skipped chs—15 dc.

Rnd 2:
Ch 3 (counts as a dc on this and following rnds), 2 dc in next dc; dc in next dc, ch 2; * dc in next dc, 2 dc in next dc; dc in next dc, ch 2; rep from * 3 times more; join in 3rd ch of beg ch-3—five 4-dc groups.

Rnd 3:
Ch 3, dc in same ch as joining; dc in next 2 dc, 2 dc in next dc; in next ch-2 sp work (dc, ch 2, dc); * 2 dc in next dc; dc in next 2 dc, 2 dc in next dc; in next ch-2 sp work (dc, ch 2, dc); rep from * 3 times more; join in 3rd ch of beg ch-3.

Rnd 4:
Ch 3, dc in next 5 dc, 2 dc in next dc; in next ch-2 sp work (dc, ch 3, dc); 2 dc in next dc; * dc in next 6 dc, 2 dc in next dc; in next ch-2 sp work (dc, ch 3, dc); 2 dc in next dc; rep from * 3 times more; join in 3rd ch of beg ch-3.

Rnd 5:
Ch 3, dc in next 7 dc, 2 dc in next dc; * † in next ch-3 sp work (dc, ch 3, dc); 2 dc in next dc †; dc in next 10 dc, 2 dc in next dc; rep from * 3 times more, then rep from † to † once; dc in next 2 dc; join in 3rd ch of beg ch-3.

Rnd 6:
Sl st in next 2 dc, ch 3, dc in same dc; ch 2, 2 dc in next dc—beg shell made; sk next 2 dc; * 2 dc in next dc; ch 2, 2 dc in next dc—shell made; sk next 2 dc; rep from * 18 times more; join in 3rd ch of beg ch-3—20 shells.

Rnd 7:
Sl st in next dc and in next ch-2 sp, ch 3, in same sp work (dc, ch 2, 2 dc)—beg shell made; ch 2; * in ch-2 sp of next shell work (2 dc, ch 2, 2 dc)—shell made; ch 2; rep from * 18 times more; join in 3rd ch of beg ch-3.

Rnd 8:
Ch 1, sc in same ch as joining; sc in each dc and 2 sc in each ch-2 sp; join in first sc—160 sc. Finish off.

Sides and Bottom Joining

Make lp on hook, join with an sc in 10th sc to right of joining on front piece; ch 2, hold front and back with wrong sides tog; carefully matching sts, sc in corresponding sc on back; * ch 2; on front, sk next 4 sc, sc in next sc, ch 2; on back, sk next 4 sc, sc in next sc; rep from * 23 times more.

Top Ruffle

Rnd 1:
Working around top opening, 3 sc in sp formed by last joining ch-2; sc in next 39 sc on front, 3 sc in sp formed by next joining ch-2; sc in next 39 sc of back; join in first sc—84 sc.

Rnd 2:
Ch 3, dc in next sc, ch 1, sk next sc; * dc in next 2 sc, ch 1, sk next sc; rep from * around; join in 3rd ch of beg ch-3—28 ch-1 sps.

Rnd 3:
Ch 1, sc in same ch as joining, in each dc, and in each ch-1 sp; join in first sc—84 sc.

Rnd 4:
Ch 3, dc in same sc; ch 2, 2 dc in next sc; ch 3, sk next 2 sc; * 2 dc in next sc; ch 2, 2 dc in next sc; ch 3, sk next 3 sc; rep from * 15 times more; join in 3rd ch of beg ch-3.

Rnd 5:
Sl st in next dc and in next ch-2 sp, beg shell in same sp; ch 3; * shell in next shell; ch 3; rep from * 15 times more; join in 3rd ch of beg ch-3.

Finish off and weave in all ends.

Rope

Ch 200; sl st in 2nd ch from hook and in each rem ch.

Finish off and weave in ends.

Finishing

Weave rope through ch-1 sps of Rnd 2 of top ruffle. Tie ends in knots.

Cherry Pie

Size:
About 6" x 4 1/2"

Materials:
Bedspread-weight crochet cotton, 100 yds white;
20 yds each green and red
Size 6 steel crochet hook, or size required for gauge
Size 18 tapestry needle

Gauge:
8 dc = 1"

Instructions

Back

Starting at center with white, ch 4.

Rnd 1 (right side):
Dc in 4th ch from hook, ch 1, in same ch work (2 dc, ch 1) 5 times; join in 3rd ch of beg 3 skipped chs—6 ch-1 sps.

Rnd 2:
Sl st in next dc and in next ch-1 sp, ch 3 (counts as a dc on this and following rnds), in same sp work (dc, ch 1, 2 dc)—beg shell made; ch 1; * in next ch-1 sp work (2 dc, ch 1, 2 dc)—shell made; ch 1; rep from * 4 times more; join in 3rd ch of beg ch-3—6 shells.

Rnd 3:
Sl st in next dc and in next ch-1 sp, beg shell in same sp; ch 1, 2 dc in next ch-1 sp; ch 1; * shell in next ch-1 sp; ch 1, 2 dc in next ch-1 sp; ch 1; rep from * 4 times more; join in 3rd ch of beg ch-3.

Rnd 4:
Sl st in next dc and in next ch-1 sp, beg shell in same sp; ch 1, (2 dc in next ch-1 sp, ch 1) twice; * shell in next ch-1 sp; ch 1, (2 dc in next ch-1 sp, ch 1) twice; rep from * 4 times more; join in 3rd ch of beg ch-3.

Rnd 5:
Sl st in next dc and in next ch-1 sp, beg shell in same sp; ch 1, (2 dc in next ch-1 sp, ch 1) 3 times; * shell in next ch-1 sp; ch 1, (2 dc in next ch-1 sp, ch 1) 3 times; rep from * 4 times more; join in 3rd ch of beg ch-3.

Rnd 6:
Sl st in next dc and in next ch-1 sp, beg shell in same sp; ch 1, (2 dc in next ch-1 sp, ch 1) 4 times; * shell in next ch-1 sp; (2 dc in next ch-1 sp, ch 1) 4 times; rep from * 4 times more; join in 3rd ch of beg ch-3.

Rnd 7:
Sl st in next dc and in next ch-1 sp, beg shell in same sp; ch 1, (2 dc in next ch-1 sp, ch 1) 5 times; * shell in next ch-1 sp; (2 dc in next ch-1 sp, ch 1) 5 times; rep from * 4 times more; join in 3rd ch of beg ch-3.

Rnd 8:
Sl st in next dc and in next ch-1 sp, beg shell in same sp; ch 1, (2 dc in next ch-1 sp, ch 1) 6 times; * shell in next ch-1 sp; ch 1, (2 dc in next ch-1 sp, ch 1) 6 times; rep from * 4 times more; join in 3rd ch of beg ch-3.

Rnd 9:
Ch 1, sc in same ch as joining, in each dc, and in each ch-1 sp; join in first sc—144 sc. Finish off.

Front

Starting at center with white, ch 4.

Rnds 1 through 9:
Rep Rnds 1 through 9 of back. At end of Rnd 9, do not finish off.

Joining

Hold back and front with wrong sides tog; ch 1, sc in same sc as joining; ch 2, sc in corresponding sc of back, ch 2; * on front, sk next 2 sc, sc in next sc, ch 2; on back, sk next 2 sc, sc in next sc; rep from * 32 times more, leaving rem 45 sc on each piece for top opening.

Top Ruffle

Rnd 1:
Ch 1, sc in same sc; 3 sc in next ch-2 sp; sc in next 45 sc, 3 sc in next ch-2 sp; sc in next 44 sc; join in first sc—96 sc.

Rnd 2:
Ch 4 (counts as a dc and a ch-1 sp), sk next sc; * dc in next sc, ch 1, sk next sc; rep from * 46 times more; join in 3rd ch of beg ch-4.

Rnd 3:
Ch 1, sc in same ch as joining, in each ch-1 sp, and in each dc; join in first sc.

Rnd 4:
Beg shell in same sc; * sk next 3 sc; shell in next sc; rep from * 23 times more; sk next dc; join in 3rd ch of beg ch-3.

Rnd 5:
Sl st in next dc and in next ch-1 sp, beg shell in same sp; shell in ch-1 sp of each shell; join in 3rd ch of beg ch-3.

Rnd 6:
Rep Rnd 5.

Finish off and weave in all ends.

Rope

With white, ch 200; sl st in 2nd ch from hook and in each rem ch.

Finish off and weave in ends.

Leaves and Cherries

First Leaf:
With green, ch 7, 3 hdc in 2nd ch from hook; (2 dc in next ch, dc in next ch) twice; in next ch work (dc, 3 trc, ch 3, sl st in last trc made, 3 trc, dc); working on opposite side in unused lps of beg ch, (dc in next lp, 2 dc in next lp) twice; 2 hdc in next lp; join in first hdc. Do not finish off.

Strands:
Ch 16; sl st in 2nd ch from hook and in next 14 chs, sl st in same hdc as joining; ch 19, sl st in 2nd ch from hook and in next 17 chs, sl st in same hdc as last sl st made; ch 16; sl st in 2nd ch from hook and in next 14 chs. Do not finish off.

Second Leaf:
Ch 12; sl st in 3rd ch from hook; sk next 3 chs, in next ch work (3 trc, dc); (2 dc in next ch, dc in next ch) twice; 5 hdc in next ch; working on opposite side in unused lps of beg ch, dc in next lp, (2 dc in next ch, dc in next ch) twice; 3 trc in next ch; join in same ch as first sl st made. Finish off, leaving a 12" end for sewing.

Cherries (make 3)**:**

With red, ch 4.

Rnd 1:
11 dc in 4th ch from hook (beg 3 skipped chs count as a dc); join in 3rd ch of beg 3 skipped chs—12 dc.

Rnd 2:
Ch 1, sc in same ch as joining and in each dc; join in first sc.

Rnd 3:
Ch 2, keeping last lp of each dc on hook, dc in next 2 sc, YO and draw through all 3 lps on hook; * keeping last lp of each dc on hook, dc in next 3 sc, YO and draw through all 4 lps on hook; rep from * twice more; join in first st. Finish off, leaving a 12" end for sewing.

Finishing

Step 1:
Referring to photo for placement, with tapestry needle, sew cherries to ends of stems. Sew leaves to center front of sachet.

Step 2:
Weave rope through sps of Rnd 2 of top ruffle. Tie ends in knots.

Sandalwood

Size:
About 6 1/4" x 3 1/2"

Materials:
Bedspread-weight crochet cotton, 100 yds cream;
20 yds each, teal, peach, and brown
Size 6 steel crochet hook, or size required for gauge
Size 18 tapestry needle

Gauge:
8 dc = 1"

Instructions

With cream, ch 36.

Row 1 (right side)**:**
Dc in 4th ch from hook (beg 3 skipped chs count as a dc), ch 2, sk next 2 chs, dc in next 2 chs; * ch 2, sk next 2 chs, dc in next 2 chs; rep from * 6 times more—nine 2-dc groups. Ch 1, turn.

Row 2:
Sc in first 2 dc, ch 2, (sc in next 2 dc, ch 2) 7 times; sc in next dc and in 3rd ch of beg 3 skipped chs. Ch 3 (counts as first dc on following rows), turn.

Row 3:
Dc in next sc, (ch 2, dc in next 2 sc) 8 times. Ch 1, turn.

Row 4:
Sc in first 2 dc, ch 2, (sc in next 2 dc, ch 2) 7 times; sc in next dc and in 3rd ch of turning ch-3. Ch 3, turn.

Rows 5 through 50:
Rep Rows 3 and 4 twenty-three times.

Row 51:
Rep Row 3. At end of Row 51, do not ch 3; do not finish off.

continued

Joining

Fold piece in half lengthwise with wrong sides tog (Rows 1 through 25 become back of piece). Ch 1, working along side in sps formed by edge dc and turning ch-3, sc in Row 51, ch 2, sc in Row 1; † ch 2, sc in next row on front, ch 2, sc in next row on back †; rep from † to † 11 times more; finish off; turn; join cream in Row 25 on opposite side, ch 1, sc in same row; ch 2, sc in Row 27; †† ch 2, sc in next row on back, ch 2, sc in next row on front ††; rep from †† to †† 11 times more. Do not finish off.

Top Ruffle

Rnd 1:
Ch 1, working across Row 51, (sc in next 2 dc, 2 sc in next ch-2 sp) 8 times; sc in next 2 dc, 3 sc in sp formed by joining ch-2; working in unused lps of beg ch and in sps formed by skipped chs, (sc in next 2 lps, 2 sc in next sp) 8 times; sc in next lp and in first sc of beg 3 skipped chs of Row 1; 3 sc in sp formed by joining ch-2; join in first sc—74 sc.

Rnd 2:
Ch 4 (counts as a dc and a ch-1 sp), sk next sc; * dc in next sc, ch 1, sk next sc; rep from * 35 times more; join in 3rd ch of beg ch-4.

Rnd 3:
Ch 1, sc in same ch as joining, in each ch-1 sp, and in each dc; join in first sc—74 sc.

Rnd 4:
Ch 3 (counts as a dc on this and following rnd), dc in same sc; ch 2, sk next sc; * 2 dc in next sc; ch 2, sk next sc; rep from * 35 times more; join in 3rd ch of beg ch-3.

Rnd 5:
Sl st in next dc and in next ch-2 sp, ch 3, dc in same sp; (ch 2, 2 dc in next ch-2 sp) 36 times; ch 2; join in 3rd ch of beg ch-3.

Finish off and weave in all ends.

Rope

With teal, ch 200; sl st in 2nd ch from hook and in each rem ch.

Finish off and weave in ends.

Diamond

With cream, ch 4.

Row 1:
Sc in 2nd ch from hook and in next 2 chs—3 sc. Ch 1, turn.

Row 2:
Sc in each sc. Ch 1, turn.

Row 3:
Sc in each sc. Change to brown by drawing lp through; cut cream. Ch 1, turn.

Row 4:
3 sc in first sc; sc in next sc, 3 sc in next sc; sc in side of edge sc on Row 2, working in unused lps of beg ch, 3 sc in next lp; sc in next lp, 3 sc in next lp; sc in side of edge sc on Row 2; join in first sc. Change to peach by drawing lp through; cut brown.

Note: *Remainder of diamond is worked in rnds. Do not turn.*

Rnd 1:
Ch 1, sc in same sc; 3 sc in next sc; * sc in next 3 sc, 3 sc in next sc; rep from * twice more; sc in next 2 sc; join in first sc. Change to teal by drawing lp through; cut peach.

Rnd 2:
Ch 1, sc in same sc and in next sc; 3 sc in next sc; * sc in next 5 sc, 3 sc in next sc; rep from * twice more; sc in next 3 sc; join in first sc. Change to cream by drawing lp through; cut teal.

Rnd 3:
Ch 1, sc in same sc and in next 2 sc; 3 sc in next sc; * sc in next 7 sc, 3 sc in next sc; rep from * twice more; sc in next 4 sc; join in first sc. Change to brown by drawing lp through; cut cream.

Rnd 4:
Ch 1, sc in same sc and in next 3 sc; 3 sc in next sc; * sc in next 9 sc, 3 sc in next sc; rep from * twice more; sc in next 5 sc; join in first sc. Change to peach by drawing lp through; cut brown.

Rnd 5:
Ch 1, sc in same sc and in next 4 sc; 3 sc in next sc; * sc in next 11 sc, 3 sc in next sc; rep from * twice more; sc in next 6 sc; join in first sc. Change to teal by drawing lp through; cut peach.

Rnd 6:
Ch 1, sc in same sc and in next 5 sc; 3 sc in next sc; * sc in next 13 sc, 3 sc in next sc; rep from * twice more; sc in next 7 sc; join in first sc. Change to cream by drawing lp through; cut teal.

Rnd 7:
Ch 1, reverse sc (see Stitch Guide on page 1) in each sc; join in first reverse sc. Finish off, leaving a 24" end for sewing.

Finishing

Step 1:
Referring to photo for placement, with tapestry needle, sew diamond to front of sachet.

Step 2:
Weave rope through ch-1 sps of Rnd 2 of top ruffle. Tie ends in knots.

Daffodil Dreams

Size:
About 6" x 4"

Materials:
Bedspread-weight crochet cotton, 100 yds white; 20 yds each, dk green and variegated yellows
Size 6 steel crochet hook, or size required for gauge
Size 18 tapestry needle

Gauge:
8 dc = 1"

Pattern Stitches

Cluster (CL):
Keeping last lp of each dc on hook, 3 dc in ch or ch-1 sp indicated; YO and draw through all 4 lps on hook—CL made.

Double Triple Crochet (dtrc):
YO 3 times, draw up lp in st indicated, (YO, draw through 2 lps on hook) 4 times—dtrc made.

Instructions

With white, ch 47.

Row 1 (wrong side):
CL (see Pattern Stitches) in 4th ch from hook (beg 3 skipped chs count as a dc); * ch 1, sk next 2 chs, CL in next ch; rep from * 13 times more; dc in next ch—15 CLs. Ch 4 (counts as first dc and ch-1 sp on following rows), turn.

Row 2 (right side):
* CL in next ch-1 sp; ch 1; rep from * 13 times more; dc in 3rd ch of beg 3 skipped chs—14 CLs. Ch 3 (counts as first dc on following rows), turn.

Row 3:
* CL in next ch-1 sp; ch 1; rep from * 13 times more; CL in sp formed by turning ch; dc in 3rd ch of turning ch-4—15 CLs. Ch 4, turn.

Row 4:
* CL in next ch-1 sp; ch 1; rep from * 13 times more; dc in 3rd ch of turning ch-3—14 CLs. Ch 3, turn.

Rows 5 through 34:
Rep Rows 3 and 4 fifteen times. At end of Row 34, do not ch 3; do not turn.

Joining

Fold piece in half lengthwise with wrong sides tog (Rows 1 through 17 become back of piece). Ch 1, working in sps formed by edge dc and turning chs, sc in Row 34, ch 3, sc in Row 1; † ch 3, sc in next row on front, ch 3, sc in next row on back †; rep from † to † 15 times more; finish off; join white on opposite side in sp formed by edge dc of Row 18, ch 1, sc in same sp; working in sps formed by edge dc and turning chs, sc in Row 17; †† ch 3, sc in next row on front, ch 3, sc in next row on back ††; rep from †† to †† 15 times more. Ch 1, turn.

Top Ruffle

Rnd 1:
3 sc in sp formed by joining ch-3; working across Row 34, sc in 3rd ch of turning ch-4, 2 sc in next ch-1 sp; (sc in next CL and in next ch-1 sp) 13 times; sc in next CL, 2 sc in next ch-1 sp; sc in next dc, 3 sc in sp formed by next joining ch-3; sc in 3rd ch of beg 3 skipped chs, working in unused lps of beg ch-47 and in sps formed by unused chs, 2 sc in next lp; (sc in next sp and in next lp) 13 times; sc in next sp, 2 sc in next lp; sc in next lp; join in first sc—72 sc.

Rnd 2:
Ch 4 (counts as a dc and a ch-1 sp), sk next sc; * dc in next sc, ch 1, sk next sc; rep from * 34 times more; join in 3rd ch of beg ch-4—36 ch-1 sps.

Rnd 3:
Ch 1, sc in same ch as joining, in each ch-1 sp, and in each dc; join in first sc—72 sc.

Rnd 4:
Ch 3, CL in next sc; * ch 3, sc in next sc, ch 3, CL in next sc; rep from * 34 times more; ch 1; to join, dc in joining sl st of Rnd 3.

Rnd 5:
Ch 1, sc in same dc; * ch 3, sc in next ch-3 sp; rep from * around; join in first sc.

Finish off and weave in ends.

Rope

With white, ch 200; sl st in 2nd ch from hook and in each rem ch.

Finish off and weave in ends.

continued

Daffodil

With variegated, ch 3.

Rnd 1:
11 hdc in 3rd ch from hook (beg 2 skipped chs count as an hdc); join in 2nd ch of beg 2 skipped chs—12 hdc.

Rnd 2:
Ch 2 (counts as an hdc on this and following rnds), working in BLs only, hdc in each hdc; join in 2nd ch of beg ch-2.

Rnd 3:
Ch 2, hdc in each hdc; join in 2nd ch of beg ch-2.

Rnd 4:
Rep Rnd 3.

Rnd 5:
Ch 1, sc in same ch as joining; ch 2; * sc in next hdc, ch 2; rep from * around; join in first sc. Finish off.

Petals:
Turn piece upside down with Rnd 1 facing you; join variegated in unused lp of any hdc on Rnd 1.

Rnd 1:
Ch 1, sc in same lp; working in rem unused lps, ch 2, sk next lp; * sc in next lp, ch 2, sk next lp; rep from * 4 times more; join in first sc—6 ch-2 sps.

Rnd 2:
* Sl st in next ch-2 sp, ch 2, in same sp work (dc, 3 trc, dc, hdc); rep from * 5 times more; join in 2nd ch of beg ch-2.

Rnd 3:
Ch 2; * † dc in next st, in next st work (dc, trc); 2 trc in next st; in next st work (trc, dc); dc in next st †; hdc in next 2 sts; rep from * 4 times more, then rep from † to † once; hdc in next st; join in 2nd ch of beg ch-2.

Finish off and weave in all ends.

Leaves

Rnd 1:
Hold daffodil upside down with Rnd 1 facing you; with green, make lp on hook, join with an sc in unused lp of any skipped hdc on Rnd 1 of daffodil; * † ch 5, working in unused lps of skipped hdc on Rnd 1, sk next lp †; sc in next lp; rep from * once more, then rep from † to † once; join in first sc.

Rnd 2:
Sl st in next ch-5 sp, ch 3 (counts as a dc on this and following rnd), in same sp work [2 dc, 3 trc, 3 dtrc (see Pattern Stitches), 3 trc, 3 dc]; in each rem ch-5 sp work (3 dc, 3 trc, 3 dtrc, 3 trc, 3 dc); join in 3rd ch of beg ch-3.

Rnd 3:
Ch 3, dc in next 2 sts; * † 2 trc in next st; trc in next 2 sts, 2 dtrc in each of next 2 sts; ch 3, sl st in last dtrc made; dtrc in same st; 2 dtrc in next st; trc in next 2 sts, 2 trc in next st †; dc in next 6 sts; rep from * once more, then rep from † to † once; dc in next 3 sts; join in 3rd ch of beg ch-3.

Finish off, leaving a 12" end for sewing.

Finishing

Step 1:
Referring to photo for placement, with tapestry needle and long end, sew daffodil to sachet.

Step 2:
Weave rope through sps of Rnd 2 of top ruffle. Tie ends in knots.

Daisies Don't Tell

Size:
About 6 1/2" x 3 1/2"

Materials:
Bedspread-weight crochet cotton, 100 yds each, white and variegated blues; 50 yards yellow
Size 6 steel crochet hook, or size required for gauge
Size 18 tapestry needle

Gauge:
8 dc = 1"

Instructions

With variegated, ch 41.

Row 1 (right side)**:**

Sc in 11th ch from hook (beg 10 skipped chs count as a ch-4 sp, an sc, and a ch-5 sp); * ch 5, sk next 4 chs, sc in next ch; rep from * 5 times more—7 ch-5 sps. Ch 5, turn.

Row 2:

* Sc in next ch-5 sp, 5 dc in next sc; sc in next ch-5 sp, ch 5; rep from * twice more; sc in sp formed by beg 10 skipped chs. Ch 5, turn.

Row 3:

* Sc in next ch-5 sp, ch 5, sc in 3rd dc of next 5 dc group; ch 5; rep from * twice more; sc in sp formed by turning ch-5. Ch 5, turn.

Row 4:

* Sc in next ch-5 sp, ch 5, sc in next ch-5 sp, 5 dc in next sc; rep from * once more; (sc in next ch-5 sp, ch 5) twice; sc in sp formed by turning ch-5. Ch 5, turn.

Row 5:

(Sc in next ch-5 sp, ch 5) twice; * sc in 3rd dc of next 5-dc group, ch 5, sc in next ch-5 sp, ch 5; rep from * once more; sc in sp formed by turning ch-5. Ch 5, turn.

Rows 6 through 49:

Rep Rows 2 through 5 eleven times. At end of Row 49, do not ch 5; do not turn.

Joining

Fold piece in half lengthwise with wrong sides tog (Rows 1 through 25 become back of piece). Ch 1, working along side in sps formed by turning chs and sides of each sc, sc in Row 49, ch 4, sc in Row 1; † ch 4, sc in next row on front, ch 4, sc in next row on back †; rep from † to † 22 times more; finish off; turn; join variegated in Row 25, ch 1, sc in same row, ch 4, sc in Row 26; †† ch 4, sc in next row on back, ch 4, sc in next row on front ††; rep from †† to †† 22 times more. Do not finish off.

Top Ruffle

Rnd 1:

Ch 1, working across Row 49, † 3 sc in each of next 7 ch-5 sps; 3 sc in sp formed by joining ch-4 †; working in sps formed by skipped chs on beg ch, rep from † to † once; join in first sc—48 sc.

Rnd 2:

Ch 3 (counts as a dc on this and following rnds), dc in next sc, ch 1, sk next sc; * dc in next 2 sc, ch 1, sk next sc; rep from * 14 times more; join in 3rd ch of beg ch-3.

Rnd 3:

Ch 1, sc in same ch as joining, in each ch-1 sp, and in each dc; join in first sc.

Rnd 4:

Ch 3, dc in same sc; ch 2, 2 dc in next sc; ch 3, sk next 2 sc; * 2 dc in next sc; ch 2, 2 dc in next sc; ch 3, sk next 3 sc; rep from * 7 times more; 2 dc in next sc; ch 2, 2 dc in next sc; ch 3, sk next 2 sc; join in 3rd ch of beg ch-3—twenty 2-dc groups.

Rnd 5:

Sl st in next dc and in next ch-2 sp, ch 3, in same sp work (dc, ch 2, 2 dc); * ch 3, in each rem ch-2 sp work (2 dc, ch 2, 2 dc); join in 3rd ch of beg ch-3.

Finish off and weave in ends.

Daisy

Center:

With yellow, ch 4.

Rnd 1:

11 dc in 4th ch from hook (beg 3 skipped chs count as a dc); join in 3rd ch of beg 3 skipped chs—12 dc.

Rnd 2:

Ch 1, sc in same ch as joining; 2 sc in next dc; * sc in next dc, 2 sc in next sc; rep from * 4 times more; join in first sc—18 sc. Finish off.

Petals:

Join white in BL of any sc on center; * † ch 11, sc in 2nd ch from hook, hdc in next ch, dc in next 8 chs †; sl st in BL of next sc; rep from * 16 times more, then rep from † to † once; join in joining sl st. Finish off, leaving a 24" end for sewing.

Rope

With variegated, ch 200; sl st in 2nd ch from hook and in each rem ch.

Finish off and weave in ends.

Finishing

Step 1:

Referring to photo for placement, with tapestry needle, sew daisy to front of sachet.

Step 2:

Weave rope through ch-1 sps on Rnd 2 of top ruffle. Tie ends in knots.

Flirtatious Curls

Size:
About 6" x 4"

Materials:
Bedspread-weight crochet cotton, 100 yds white;
 20 yds each, pink, orange, green, lavender, and blue
Size 6 steel crochet hook, or size required for gauge
Size 18 tapestry needle
Sewing needle and matching thread

Instructions

With white, ch 28.

Row 1 (right side):
In 4th ch from hook (beg 3 skipped chs count as a dc) work (2 dc, ch 2, 2 dc)—shell made; * sk next 2 chs, in next ch work (2 dc, ch 2, 2 dc)—shell made; sk next 2 chs, dc in next ch, sk next 2 chs, shell in next ch; rep from * once more; sk next 2 chs, shell in next ch; sk next 2 chs, dc in next ch. Ch 3 (counts as first dc on following rows), turn.

Row 2:
Shell in ch-2 sp of each of next 2 shells; * dc in next dc, shell in ch-2 sp of each of next 2 shells; rep from * once more; dc in 3rd ch of beg 3 skipped chs. Ch 3, turn.

Row 3:
Shell in each of next 2 shells; * dc in next dc, shell in each of next 2 shells; rep from * once more; dc in 3rd ch of turning ch-3. Ch 3, turn.

Rows 4 through 30:
Rep Row 3. At end of Row 30, do not ch 3; do not finish off.

Joining

Fold piece in half lengthwise with wrong sides tog and Rows 1 and 30 at top (Rows 1 through 15 become back of piece). Ch 1, working in sps formed by edge dc and turning chs, sc in Row 30, ch 3, sc in Row 1; † ch 3, sc in next row on front, ch 3, sc in next row on back †; rep from † to † 13 times more; finish off; turn; join white in side of Row 15 on opposite side; ch 3, sc in same sp; sc in Row 16; †† ch 3, sc in next row on back, ch 3, sc in next row on front ††; rep from †† to †† 13 times more. Do not finish off.

Top Ruffle

Rnd 1:
Ch 1, working across Row 30, sc in each dc and in 3rd ch of turning ch-3, 4 sc in side joining; working in unused lps of beg ch and in sps formed by unused chs on opposite side, sc in each unused lp and 2 sc in each sp; 4 sc in side joining; join in first sc—62 sc.

Rnd 2:
Ch 4 (counts as a dc and a ch-1 sp), sk next sc, dc in next sc; * ch 1, sk next sc, dc in next sc; rep from * 28 times more; ch 1, sk next sc; join in 3rd ch of beg ch-4.

Rnd 3:
Ch 1, sc in same ch as joining, in each dc, and in each ch-1 sp; join in first sc—62 sc.

Rnd 4:
Ch 3 (counts as a dc on this and following rnds); * sk next 2 sc, shell in next sc; sk next 2 sc, shell in next sc; sk next 2 sc, dc in next sc; rep from * 5 times more; sk next 2 sc, shell in next sc; sk next 2 sc; shell in next sc; sk next 2 sc; join in 3rd ch of beg ch-3.

Rnd 5:
Ch 3, shell in each of next 2 shells; * dc in next dc, shell in each of next 2 shells; rep from * 7 times more; join in 3rd ch of beg ch-3.

Finish off and weave in all ends.

Rope

With white, ch 200; sl st in 2nd ch from hook and in each rem ch.

Finish off and weave in ends.

Color Curls (make 5—one each pink, orange, green, lavender, and blue)

Ch 50; 2 dc in 4th ch from hook; 3 dc in each rem ch. Finish off and weave in ends.

Finishing

Step 1:
Weave rope through sps on Rnd 2 of top ruffle. Tie ends in knots.

Step 2:
With sewing needle and matching thread, sew curls to top right-hand corner where top ruffle begins and opposite rope ends.

Hearts' Delight

Size:
About 6" x 4"

Materials:
Bedspread-weight crochet cotton, 100 yds white; 50 yds pink
Size 6 steel crochet hook, or size required for gauge
Size 18 tapestry needle

Gauge:
8 dc = 1"

Instructions

With white, ch 41.

Row 1 (wrong side)**:**
Dc in 8th ch from hook **(**beg 7 skipped chs count as a ch 2, a dc, and a ch-2 sp**)**; ***** ch 2, sk next 2 chs, dc in next ch; rep from ***** 10 times more—12 ch-2 sps. Ch 5 **(**counts as first dc and ch-2 sp on following rows**)**, turn.

Row 2 (right side)**:**
***** Dc in next dc, ch 2; rep from ***** 10 times more; sk next 2 chs of beg 7 skipped chs, dc in next ch. Ch 5, turn.

Row 3:
***** Dc in next dc, ch 2; rep from ***** 10 times more; sk next 2 chs of turning ch-5, dc in next ch. Ch 5, turn.

Rows 4 through 36:
Rep Row 3. At end of Row 36, do not ch 5; do not turn.

Side Joinings

Fold piece in half lengthwise with wrong sides tog **(**Rows 1 through 18 become back of piece**)**. Ch 1, working in sps formed by edge dc and turning chs, sc in Row 36, ch 2, sc in Row 1; † ch 2, sc in next row on front, ch 2, sc in next row on back †; rep from † to † 16 times more; finish off; turn; join white in side of Row 18 on opposite side; ch 1, sc in same sp; ch 2, sc in side of Row 19; rep from † to † 17 times.

Top Ruffle

Rnd 1:
Working across Row 36, ***** † 3 sc in each of next 2 ch-2 sps; 2 sc in next ch-2 sp; 3 sc in next ch-2 sp; 2 sc in next ch-2 sp †; rep from † to † once more; 3 sc in each of next 2 ch-2 sps; 3 sc in joining ch-2 sp; rep from ***** once more; join in first sc—70 sc.

Rnd 2:
Ch 4 **(**counts as a dc and a ch-1 sp on this and following rnds**)**, sk next sc; ***** dc in next sc, ch 1; rep from ***** 33 times more; join in 3rd ch of beg ch-4.

Rnd 3:
Ch 1, sc in same ch as joining, in each ch-1 sp, and in each dc; join in first sc.

Rnd 4:
Ch 4, dc in same sc; dc in next sc; ***** in next sc work **(**dc, ch 1, dc**)**; dc in next sc; rep from ***** 33 times more; join in 3rd ch of beg ch-4.

Rnd 5:
Sl st in next ch-1 sp, ch 4, dc in same sp; ch 1, dc in next dc, ch 1; ***** in next ch-1 sp work **(**dc, ch 1, dc**)**; ch 1, dc in next dc, ch 1; rep from ***** 33 times more; join in 3rd ch of beg ch-4.

Finish off and weave in ends.

Large Heart

With pink, ch 4.

Rnd 1:
In 4th ch from hook **(**beg 3 skipped chs count as a dc**)** work **(**6 dc, 2 hdc, sc, sl st, ch 1, sc, 2 hdc, 6 dc**)**; join in 3rd ch of beg 3 skipped chs—20 sts.

Rnd 2:
Ch 3, 2 dc in same ch as joining; dc in next 4 sts, 2 dc in each of next 3 sts; hdc in next 2 sts, sc in next st, sk next ch-1 sp, hdc in next 2 sts, 2 dc in each of next 3 sts; dc in next 4 sts; join in 3rd ch of beg ch-3—28 sts.

Rnd 3:
Ch 2, 2 hdc in same ch as joining; hdc in next 7 sts, 2 dc in each of next 4 sts; 2 hdc in next st; sc in next st, sl st in next st, ch 1, sc in next st, 2 hdc in next st; 2 dc in each of next 4 sts; hdc in next 7 sts; join in 2nd ch of beg ch-2—40 sts.

Rnd 4:
Ch 1, 3 sc in same ch as joining; sc in next 13 sts, hdc in next sc, 2 hdc in each of next 3 sts; hdc in next st, sc in next st, sl st in next sl st, ch 1, sk next ch-1 sp, sc in next st, hdc in next st, 2 hdc in each of next 3 sts; hdc in next st, sc in next 13 sts; join in first sc.

continued

Finish off, leaving an 18" end for sewing.

Small Hearts (make 4)
With pink, ch 4.

Rnd 1:
In 4th ch from hook (beg 3 skipped chs count as a dc) work (6 dc, 2 hdc, sc, sl st, ch 1, sc, 2 hdc, 6 dc); join in 3rd ch of beg 3 skipped chs—20 sts.

Rnd 2:
Ch 3 (counts as a dc), 2 dc in same ch as joining; dc in next 4 sts, 2 dc in each of next 3 sts; hdc in next st, sc in next st, sl st in next st, ch 1, sc in next st, hdc in next st, 2 dc in each of next 3 sts; dc in next 4 sts; join in 3rd ch of beg ch-3.

Finish off, leaving a 12" end for sewing.

Rope

With white, ch 200; sl st in 2nd ch from hook and in each rem ch.

Finish off and weave in ends.

Finishing

Step 1:
Referring to photo for placement, with tapestry needle, sew large heart to front of sachet.

Step 2:
Weave rope through ch-1 sps of Rnd 2 of top ruffle. Tie knots at each end of rope about 1" from end.

Step 3:
Hold 2 small hearts with wrong sides together; with tapestry needle, sew together around edges, tucking end of rope between hearts. Repeat with remaining hearts.

Lavender Flight

Size:
About 6" x 4 1/2"

Materials:
Bedspread-weight crochet cotton, 100 yds lavender; 20 yds each, dk purple and white
Size 6 steel crochet hook, or size required for gauge
Size 18 tapestry needle

Gauge:
8 dc = 1"

Instructions

With lavender, ch 39.

Row 1 (wrong side):
5 dc in 6th ch from hook—beg shell made (beg 5 skipped chs count as a dc and a ch-2 sp); sk next 2 chs, dc in next ch; * sk next 2 chs, 5 dc in next ch—shell made; sk next 2 chs, dc in next ch; rep from * 4 times more—6 shells. Ch 3 (counts as first dc on following rows), turn.

Row 2 (right side):
Working in BLs only, shell in 3rd dc of next shell; * sk next 2 dc, dc in next dc, shell in 3rd dc of next shell; rep from * 4 times more; dc in next ch of beg 5 skipped chs. Ch 3, turn.

Row 3:
Working in FLs only, shell in 3rd dc of next shell; * sk next 2 dc, dc in next dc, shell in 3rd dc of next shell; rep from * 4 times more; dc in 3rd ch of turning ch-3. Ch 3, turn.

Row 4:
Working in BLs only, shell in 3rd dc of next shell; * sk next 2 dc, dc in next dc, shell in 3rd dc of next shell; rep from * 4 times more; dc in 3rd ch of turning ch-3. Ch 3, turn.

Rows 5 through 28:
Rep Rows 3 and 4 twelve times. At end of Row 28, do not ch 3; do not turn.

Joining

Fold piece in half lengthwise with wrong sides tog (Rows 1 through 14 become back of piece). Ch 1, working in sides of each row in sps formed by edge dc and turning chs, sc in side of Row 28, ch 3, sc in side of Row 1; † ch 3, sc in side of next row on front, ch 3, sc in side of next row on back †; rep from † to † 12 times more; finish off; join lavender in sp made by turning ch-3 on opposite side of Row 14, ch 1, sc in same sp; ch 3, working in sides of each row in sps formed by edge dc and turning chs, sc in side of Row 15; †† ch 3, sc in side of next row on back, ch 3, sc in side of next row on front ††; rep from †† to †† 12 times more. Do not finish off.

Top Ruffle

Rnd 1:
Ch 1, working in unused lps of Row 28, sc in next 37 dc, 3 sc in next ch-3 sp, working along beg ch, sc in each unused lp and in each unworked ch of next 37 chs; 3 sc in next ch-3 sp; join in first sc—80 sc.

Rnd 2:
Ch 4 (counts as a dc and a ch-1 sp), sk next sc; * dc in next dc, ch 1, sk next sc; rep from * 38 times more; join in 3rd ch of beg ch-4.

Rnd 3:
Ch 1, sc in same ch as joining, in each ch-1 sp, and in each dc; join in first sc. Change to dk purple by drawing lp through; cut lavender.

Rnd 4:
Beg shell in same sc; sk next sc, dc in next sc, sk next 2 sc; * shell in next sc; sk next sc, dc in next sc, sk next 2 sc; rep from * 14 times more; join in 3rd ch of beg ch-3—16 shells.

Rnd 5:
Sl st in next 2 dc, beg shell in same dc as last sl st made; sk next 2 dc, dc in next dc; * shell in 3rd dc of next shell; sk next 2 dc, dc in next dc; rep from * 14 times more; join in 3rd ch of beg ch-3. Change to white by drawing lp through; cut dk purple.

Rnd 6:
Ch 1, sc in same ch as joining; ch 1; * sc in next dc, ch 1; rep from * around; join in first sc. Finish off.

Side Shells

Hold sachet with right side facing you and one side edge at top; join dk purple in first sp formed by joining ch-3 along one side; ch 3, 6 dc in same sp; * sk next ch-3 sp, sc in next ch-3 sp, sk next ch-3 sp, 7 dc in next sp; rep from * 5 times more. Finish off.

Beginning at bottom of sachet, work other side in same manner.

Weave in all ends.

Rope

With dk purple, ch 200; sc in 2nd ch from hook and in each rem ch. Finish off and weave in ends.

Butterfly

With white, ch 4.

Rnd 1:
In 4th ch from hook work (dc, ch 1) 7 times (beg 3 skipped chs count as a dc and a ch-1 sp); join in 3rd ch of beg ch-4.

Rnd 2:
Sl st in next ch-1 sp, ch 4, dc in same sp; in each rem ch-1 sp work (dc, ch 1, dc); join in 3rd ch of beg ch-4.

Rnd 3:
Sl st in next ch-1 sp, ch 3, in same sp work (dc, ch 1, 2 dc); in each rem ch-1 sp work (2 dc, ch 1, 2 dc); join in 3rd ch of beg ch-3.

Rnd 4:
Sl st in next dc and in next ch-1 sp, ch 3, in same sp work (dc, ch 2, 2 dc); in each rem ch-1 sp work (2 dc, ch 2, 2 dc); join in 3rd ch of beg ch-3. Finish off.

Join dk purple in any ch-2 sp.

Rnd 5:
Ch 3, 6 dc in same sp; sk next dc, sc between next 2 dc; * 7 dc in next ch-2 sp; sk next dc, sc between next 2 dc; rep from * 6 times more; join in 3rd ch of beg ch-3. Finish off.

With dk purple, and leaving a 12" end at beg, ch 15. Finish off, leaving a 12" end.

Finishing

Step 1:
Fold butterfly in half, having four 7-dc groups on each side. Tie ch-15 around center of folded butterfly.

Step 2:
Referring to photo for placement, with tapestry needle, sew butterfly to front of sachet.

Step 3:
Weave rope through ch-1 sps of Rnd 2 of top ruffle. Tie ends in knots.

Pine Holiday

Size:
About 6" x 4 1/2"

Materials:
Bedspread-weight crochet cotton, 50 yds white;
25 yds each, red and green
Size 6 steel crochet hook, or size required for gauge

Gauge:
8 dc = 1"

Instructions

Note: *Carry colors not in use up side of work. Do not finish off or rejoin.*

With white, ch 32.

Row 1 (wrong side)**:**
Sc in 2nd ch from hook and in each rem ch—31 sc. Ch 3 **(**counts as first dc on following rows**)**, turn.

Row 2 (right side)**:**
Sk next 2 sc, 5 dc in next sc; * sk next 3 sc, 3 dc in next sc; sk next 3 sc, 5 dc in next sc; rep from * twice more; sk next 2 sc, dc in next sc—31 dc. Change to red by drawing lp through; ch 1, turn.

Row 3:
Sc in each dc and in 3rd ch of turning ch-3—31 sc. Change to green by drawing lp through; ch 1, turn.

Row 4:
Working in BLs only, sc in each sc. Change to white by drawing lp through; ch 3, turn.

Row 5:
Sk next 2 sc, 3 dc in next sc; * sk next 3 sc, 5 dc in next sc; sk next 3 sc, 3 dc in next sc; rep from * twice more; sk next 2 sc, dc in next sc—29 dc. Change to red by drawing lp through; ch 1, turn.

Row 6:
Sc in each dc and in 3rd ch of turning ch-3—29 sc. Change to green by drawing lp through; ch 1, turn.

Row 7:
Working in FLs only, sc in each sc. Change to white by drawing lp through; ch 3, turn.

Row 8:
Sk next sc, 5 dc in next sc; * sk next 3 sc, 3 dc in next sc; sk next 3 sc, 5 dc in next sc; rep from * twice more; sk next sc, dc in next sc—31 dc. Change to red by drawing lp through; ch 1, turn.

Row 9:
Sc in each dc and in 3rd ch of turning ch-3—31 sc. Change to green by drawing lp through; ch 1, turn.

Rows 10 through 63:
Rep Rows 4 through 9 nine times.

Rows 64 through 68:
Rep Rows 4 through 8. At end of Row 68, do not change to red; cut red and green. With white, ch 1, turn.

Row 69:
Sc in each dc and in 3rd ch of turning ch-3. Ch 1, turn.

Row 70:
Sc in each sc—31 sc.

Side Joinings

Fold piece in half lengthwise with wrong sides tog **(**Rows 1 through 35 become back of piece**)**. Ch 1, working along side, sc in side of edge sc on Row 70, ch 2, sc in side of edge sc on Row 1; working only in sps formed by edge dc and turning chs, † ch 2, sc in next sp on front, ch 2, sc in next sp on back †; rep from † to † 10 times more; finish off; turn; join white in side of Row 36 on front; ch 1, sc in same row, ch 2, sc in corresponding sc row on back; working only in sps formed by edge dc and turning chs, †† ch 2, sc in next sp on front, ch 2, sc in next sp on back ††; rep from †† to †† 10 times more; ch 2, sc in side of Row 70, ch 2, sc in side of Row 1. Do not finish off.

Top Ruffle

Rnd 1:
Ch 1, 3 sc in sp formed by last joining ch-2; working across Row 70, sc in next 31 sc, 3 sc in next joining ch-2 sp; working in unused lps of beg ch, sc in next 31 lps; join in first sc—68 sc.

Rnd 2:
Ch 4 **(**counts as a dc and a ch-1 sp**)**, sk next sc; * dc in next sc, ch 1, sk next sc; rep from * 32 times more; join in 3rd ch of beg ch-4. Change to red by drawing lp through; do not cut white.

Rnd 3:
Ch 1, sc in same ch as joining, in each ch-1 sp, and in each dc; join in first sc—68 sc. Change to white by drawing lp through.

Rnd 4:
Ch 3, 2 dc in same sc; sk next sc; * 3 dc in next sc; sk next 2 sc; rep from * 21 times more; join in 3rd ch of beg ch-3—66 dc. Change to red by drawing lp through; cut white.

Rnd 5:
Ch 1, sc in same ch as joining and in each dc; join in first sc. Change to green by drawing lp through; cut red.

Rnd 6:
Ch 1, sc in same sc; ch 3, sk next sc; * sc in next sc, ch 3, sk next sc; rep from * 31 times more; join in first sc.

Finish off and weave in all ends.

Rope
With green, ch 200; sl st in 2nd ch from hook and in each rem ch.

Finish off and weave in ends.

Finishing
Weave rope through ch-1 sps on Rnd 2 of top ruffle. Tie ends in knots.

Rose Secret

Size:
About 6" x 4"

Materials:
Bedspread-weight crochet cotton, 100 yds white; 20 yds each, burgandy, rose and green
Size 6 steel crochet hook, or size required for gauge
Size 18 tapestry needle

Gauge:
8 dc = 1"

Instructions
With white, ch 36.

Row 1 (wrong side)**:**
Dc in 5th ch from hook and in 4th ch from hook—beg cross st made; * sk next ch, dc in next ch, dc in skipped ch—cross st made; rep from * 14 times more; dc in next ch—16 cross sts. Ch 3 (counts as first dc on following rows), turn.

Row 2 (right side)**:**
Sk first 2 dc, dc in next dc, dc in last dc skipped—cross st in cross st made; * sk next dc, dc in next dc, dc in skipped dc—cross st in cross st made; rep from * 14 times more; dc in 3rd ch of beg 3 skipped chs. Ch 3, turn.

Row 3:
Sk first dc, cross st over each cross st; dc in 3rd ch of turning ch-3. Ch 3, turn.

Rows 4 through 34:
Rep Row 3. At end of Row 34, do not ch 3; do not turn.

Side Joinings
Fold piece in half lengthwise with wrong sides tog (Rows 1 through 16 become back of piece)). Ch 1, working in sps formed by edge dc and turning chs, sc in Row 34, ch 3, sc in Row 1; † ch 3, sc in next row on front, ch 3, sc in next row on back †; rep from † to † 14 times more; finish off; turn; join white in side of Row 1 on opposite side; ch 1, sc in same sp; sc in Row 34; †† ch 3, sc in next row on back, ch 3, sc in next row on front ††; rep from †† to †† 14 times more. Finish off.

Top Ruffle
Rnd 1:
Hold sachet with opening at top and back (Row 1) facing you; join white in first skipped ch of beg ch-36; ch 1, sc in same ch; working in unused lps of beg ch, sc in next 33 lps, 3 sc in sp formed by side joining; working along front on Row 34, sc in next 33 dc and in 3rd ch of turning ch-3, 3 sc in sp formed by side joining; join in first sc—74 sc.

Rnd 2:
Ch 4 (counts as a dc and a ch-1 sp); * sk next sc, dc in next sc, ch 1; rep from * 35 times more; join in 3rd ch of beg ch-4—37 ch-1 sps.

Rnd 3:
Ch 1, sc in same ch as joining, in each dc, and in each ch-1 sp; join in first sc—74 sc.

Note: *Rnds 4 through 7 are worked in continuous rnds. Do not join; mark beg of rnds.*

continued

Rnd 4:
Ch 1, sc in same sc; * ch 3, sk next sc, sc in next sc; rep from * 35 times more.

Rnd 5:
* Sc in next ch-3 sp, ch 3; rep from * around.

Rnds 6 and 7:
Rep Rnd 5.

Rnd 8:
* Sc in next ch-3 sp, ch 3; rep from * around; join in first sc.

Finish off and weave in all ends.

Rope
With white, ch 200; sl st in 2nd ch from hook and in each rem ch.

Finish off and weave in ends.

Rose
With burgandy, ch 53.

Row 1:
Sc in 9th ch from hook (beg 8 skipped chs count as an sc, a ch-5 sp, and an sc); * ch 5, sk next 3 chs, sc in next ch; rep from * 10 times more—12 ch-5 sps. Ch 1, turn.

Row 2:
5 sc in each of next 2 ch-5 sps; 7 hdc in each of next 3 ch-5 sps; [sl st in next ch-5 sp, in same sp work (9 dc, sl st)] 4 times; [sl st in next ch-5 sp, ch 3, in same sp work (12 trc, sl st)] 3 times. Finish off. Turn.

Row 3:
With pink, make lp on hook and join with an sc in first sl st; (sc in next 12 trc, in 3rd ch of ch-3, and in next sl st) 3 times; (sc in next sl st, in next 9 dc, and in next sl st) 4 times; sl st in each rem st. Finish off, leaving a 12" end for sewing.

Roll piece to form rose; tie thread ends tog to secure.

Leaves (make 2)
With green, ch 15.

Rnd 1:
Dc in 4th ch from hook (beg 3 skipped chs count as a dc) and in next 10 chs, 7 dc in next ch; working along opposite side in unused lps of beg ch, dc in next 10 lps, 5 dc in next lp; join in 3rd ch of beg 3 skipped chs—34 dc.

Rnd 2:
Ch 1, sc in same ch as joining and in next 4 dc; hdc in next 5 dc, dc in next 2 dc, 2 dc in each of next 3 dc; 3 trc in next dc; ch 3, sl st in last trc made; 2 trc in next dc; 2 dc in next dc; dc in next 2 dc, hdc in next 5 dc, sc in next 9 dc; join in first sc.

Finish off, leaving a 12" end for sewing.

Finishing
Step 1:
Referring to photo for placement, with tapestry needle, sew leaves and rose to front of sachet.

Step 2:
Weave rope through sps on Rnd 2 of top ruffle. Tie ends in knots.

Sweet Rosebud

Size:
About 5 1/2" x 3 1/4"

Materials:
Bedspread-weight crochet cotton, 100 yds variegated pinks; 20 yds each, white and green
Size 6 steel crochet hook, or size required for gauge
Size 18 tapestry needle

Instructions
With variegated, ch 36.

Row 1 (wrong side)**:**
Sc in 6th ch from hook (5 skipped chs count as an sc and a ch-3 sp), ch 3, sk next 2 chs; * sc in next ch, ch 3, sk next 2 chs; rep from * 8 times more; sc in next ch—11 ch-3 sps. Ch 3, turn.

Row 2 (right side)**:**
* Sc in next ch-3 sp, ch 3; rep from * 9 times more; sc in sp formed by beg 5 skipped chs. Ch 3, turn.

Row 3:
* Sc in next ch-3 sp, ch 3; rep from * 9 times more; sc in sp formed by turning ch-3. Ch 3, turn.